Natural Resource Management:
Economic, Political, and Social Forces
Influence on Environmental Policies

A Collection of Essays

Natural Resource Management: Economic, Political, and Social Forces Influence on Environmental Policies

A Collection of Essays

Amy Bradley, MNR

Viridescence.org
Oakridge, Oregon

Copyright © 2018 Amy Bradley

All rights reserved. No part of this book may be reproduced in any form or by any electronic or mechanical means, including information storage and retrieval systems, without permission in writing from the publisher, except by reviewers, who may quote brief passages in a review.

ISBN: 9781792012587

Library of Congress Control Number

Editing by Amy Bradley
Front cover image and photographs by Amy Bradley

Printed and bound in the United States of America
First printing December 2018

Published by the Viridescence Organization
49627 High Prairie Loop
Oakridge, OR, USA 97463

Visit http://www.viridescence.org/.

TABLE OF CONTENTS

Introduction 7

Social Response to Environment Issues 9

Social and Cultural Influence on Environmental Policy 25

Forces That Drive Environmental Policies 39

Driving Forces of Environmental Politics and Policy 51

Environmental Policy Process Participants 65

Markets, Science, Conflict, and Sustainability's Effect on Environmental Policy 81

The American Wilderness: An Essay 94

Wilderness: The Effects of Human Expansion 97

Position Statement: Carbon Emissions 113

Environmental Social Justice 119

Glossary 125

Index 128

INTRODUCTION

Natural Resource Management: Economic, Political, and Social Forces Influence on Environmental Policies delves into the reasoning behind the decision making that going into creating environmental policies. This book examines the stakeholders involved in the development of environmental policies and their motives behind their decisions. Additionally, a breakdown of the economic, political and social factors that drive environmental policy.

"Social Response to Environmental Issues" discusses the different techniques used by social scientist to measure a citizen's orientation toward the environment as well as the rise of the modern environmental movement, the difference in rural and urban communities' views on natural resources, and the idea of an environmental hypocrite.

"Social and Cultural Influences on Environmental Policy" explores the environmental policies and problems of Nepal, Korea, the United States, and Australia and how different cultural and religious traditions effect environmental policy.

"Forces that Drive Environmental Policies" examines how environmental policy differs between post-industrial countries, post-communist countries, and developing countries. In particular, how the economic and governmental structures influence environmental policy.

"Driving Forces of Environmental Politics and Policy" focuses on the effects population growth, economic development, technological change, globalization, and changing values have on the environmental policy process.

"Environmental Policy Process Participants" explains the roles of each participant in the process of creating environmental policy. Participants includes non-governmental interest groups, industrial groups, interest groups, and governmental offices, such as President and Congress.

"Markets, Science, Conflict, and Sustainability's Effect on Environmental Policy" describes the role of each in the process of creating environmental policy.

"The American Wilderness" and "Wilderness: the Effects of Human Expansion" attempt to explain the term wilderness and how it is improperly applied to the undeveloped areas in the United States called wilderness areas. This essay reals how the American landscape has been transformed over time and what we have today is far from pristine.

"Position Statement: Carbon Emissions" discusses placing limits on carbon emissions, specifically by using the cap and trade system and the opinions of both proponents and opponents.

"Environmental Social Justice: a Case Study of Northeast Portland, Oregon" examines the risks associated with environmental social justice in Portland, their uneven distribution, sources of the dissimilarities, reasons discrimination proof is difficult, and current actions attempting to fix the imbalance in environmental equality.

Social Response to Environment Issues

Individual members of society's response to environmental issues varies depending on their social class, political affiliation, place of residence, and livelihood. Because of the variety of influences on an individual's relationship and view of natural resources, social scientist have employee a multitude of techniques in order to gauge an individual's response and as such society's response to environmental issues.

Measuring Citizen Orientation toward the Environment

The measure of the success of environment policy both conceptually and empirically is difficult to conclude because of the inconsonant data and unreliable collection methods along with variations in the interpretation of the data. Ultimately, success associates with achievement of the policy's goals. Policies regarding environmental issues are difficult to quantify due to the long-term commitments applied to social oriented values and goals. The difficultly in assessing environmental trends are in part due to inconsistent data collection and unreliable monitoring of environmental conditions.

Social scientists use a variety of quantitative and qualitative methods to study social values. Both quantitative and qualitative methods determine and comprehend people's desire and perception of natural resource issues and consider observation the best technique to ascertain social values. Quantitative approaches to the study of social values observes, measures, and analyzes opinion and attitudinal measurements, such as a survey questionnaire, that involve the use of numerical data and statistics (Steel et al., 1993). Conversely, qualitative research involves the collection of more descriptive types of data through close personal observation in order to provide a description of social values. The results are beneficial for designing better surveys that in turn produce quantitative data.

The study of social values employs gathering data through a variety of methods of which the three most commonly used are opinion and attitudinal measurement, cognitive approaches and ethnography. Other approaches include historical research, comparative studies, and network analysis.

Opinion and attitudinal measurement uses the basic assumption that the behavior of people is the result of their attitude and preferences and considers the opinions and values of its citizens (Steel et al., 1993). This method is useful for determining the success of a specific natural resource management action or policy and/or determining citizen priorities concerning the use of natural resources.

Cognitive approaches utilize citizens' attitudes from their own point of view in order to obtain the thoughts of a population and to predict its reaction to change. The cognitive approaches involve three steps beginning with an interview containing open-ended questions in order to gain individual perceptions and obtain value statements; choosing the most important value statements then sorting by importance; and finally presenting to a representative sample of the population a formal survey containing the chosen most important value statements, thus incorporating both quantitative and qualitative techniques.

Ethnography is the description of a cultural system by an outside observer in which gathering the data does not use predetermined categories. When gathering ethnographic data, social scientist do not use predetermined categories, instead allowing the setting to dictate what they observe (Steel et al., 1993). This qualitative method provides useful data illustrating how citizens interact with natural resources. In order to avoid creating a bias picture of citizen preferences, social scientist must ensure the methods of data gathering use the proper sample size or selection, survey and interview type, observation style and proper analysis of content. However, these techniques in certain situations can provide useful information but are not necessarily representative of the population.

In quota sampling, the social scientist uses demographic and economic information in order to decide on the group of interest and then collects data from the entire group in a nonrandom manner (Steel et al., 1993). When researchers have no plan of the number or divergence of participants to study and go into the field to find the informants, they are performing a purposive sampling. Another sampling technique, snow ball sampling is the technique of questioning an individual and asking them to identify other possible participants for the scientific research. This technique has the ability to identify various stockholders and their network; however, if care should be exercised as to not generalize the entire population. Lastly, hazard sampling as a method in common use but one they highly discourage due to its high probability of bias. Hazard sampling allows researchers a way to get an idea of what is happening in a community by simply grabbing anyone willing to be interviewed.

According to Steel et al. (1993), surveys, in the form of mail, telephone and personal interviews, provide accurate and representative information about social issues that is easily quantifiable and reduces the cost of contacting large number of users. Accurate and representative information results from proper random sampling. Conversely, researchers will inadvertently limit the information they receive or receive superficial information because of the wording in a question or for omitting important topics. Steel et al. (1993) suggest conducting cognitive or ethnographic research before designing a questionnaire in order to identify keys areas of concerns. Moreover, direct observation through inductive and deductive field research permits flexibility in research focus and the capability to develop an in-depth understanding of social values (Steel et al., 1993). Interviews range from informal to structured interviews depending on the amount of control the researcher exercises over the informants' response. Informal interviews involve the researcher casually speaking to individuals in the field and making notes shortly afterwards. Unstructured interviews take place between a researcher and an informant who knows the discussion is an interview and not casual conversation. Semi-structured interviews are considered the best when only one chance is available to interview the informant. These interviews have a list of questions for the researcher to ask but allow the researcher to follow leads with discretion using formal interview guides to ensure consistency of results, especially if more than one researcher is performing interviews. Structured interviews involve close-ended questions for the interviewers proving them with the most control for researchers. All participants respond to the same question ensuring the compatibility of data. Lastly, content analysis is a useful technique to ascertain social values. Researchers examine and quantify the public's written or spoken communications when there is a lack of other data gathering options. However, bias can occur if the comments come from only on constituency group.

The Rise of the Modern Environmental Movement

The rise of the modern environmental movement is the result of population, technological and environmental movement value change among citizens in the post-industrial world following World War II. Previously, the public lands of the Western United States were the source of economic wealth with timber extraction and cattle grazing lands. By the 1960s, public opinion was changing with a new focus on non-extractive values such as forest preservation and recreational parks creating a negative impact on timber and cattle grazing through new controversy and litigation (Steel, 2009).

The change in the value of natural resources, some argue, is the result of a change in the motivation for social behavior especially the younger generations higher order needs supplanting fundamental substance needs and the acquisition of materials.

Post-industrial societies experience a population shift from rural citizens moving into urban areas fueling the emergence of the New Resource Management Paradigm (Steel, 2009). There was a major shift toward urbanization created a shift in economic and political power during the nineteenth century and twentieth century. Higher education and employment are the driving forces of the migration to urban areas. Meanwhile, retirees are moving to the rural areas for recreational activities resulting in even more urban influence over rural policies (Steel, 2009).

The shift in population cores lead to a shift in employment with the service sector accounting for seventy-two percent of U.S. economy while agriculture and natural resources make up two percent (Steel, 2009). Urban areas experience less poverty and unemployment the rural peripheral. Rural residences, in turn, want to increase natural resource production to elevate the unemployment and poverty problems while urban residence fight to protect and preserve the natural resources.

Steel (2009) found technological innovation and change as central to the relationship between the urban core and the rural periphery". Environment management issues are highly complex due to technological innovations that provide an increasing amount of data. Steel uses the example of political scientist Mathew Cahn as an opponent of citizen participation in such highly technical issues as environmental management (Steel, 2009). Citizen participation is central to the democratic system and excluding it will erode its central institution, therefore allowing technical expertise to be the primary determinant of environmental policy leads to an imbalance in the system. The imbalance in the system will shift the other direction with the lack of technical expertise increasing the probability of mismanagement of complex problems.

The emergence of the two opposing natural resource management paradigm is due to a change in the value of the use of natural resources. Citizens of the post-industrial world are post-materialist as a result of the change in values from basic needs to higher order motivations leading to, especially in urban areas, development of the environmental movement. The idea of environmental protection directly conflicts with the lifestyle and economic factors of the rural peripheral giving rise to the *Dominant Resource Management Paradigm* (DRMP) and the *New Resource Management Paradigm* (NRMP).

United States policy toward environment issues reflects the increase of public concern. According to Kraft and Vig (2010), the federal government until 1970s played a limited role in environmental policy making with the exception of public land management. The affluent and well-educated members of society during the 1970s began placing new emphasis on the quality of life as did concerns for environment protection effectively placing public press on the federal government to act with more vigor to prevent environment degradation (Kraft & Vig, 2010).

The Regan presidency of the 1980s brings a markedly different environmental policy agenda to the administration reevaluating the 1970s environmental policies in response to what Kraft and Vig (2010) call Regan's "desire to reduce the scope of government regulation, shift responsibility to the states, and rely more on the private sector" (p. 13). National and grassroots environmental groups organize in response to Regan's lax enforcement of pollution laws and predevelopment resource polices (Kraft & Vig, 2010). Environmental spending continues to see rise through the administration of George H.W. Bush, and Bill Clinton again seeing reductions during the administration of George W. Bush.

During President Obama's era in office, there was an increase in environmental policies and spending. During President Obama's term in office he established the Pacific Remote Islands Marine National Monument the largest marine reserve in the world and expanded the California Coastal National Monument. He signed a ban on the use of microbeads in certain cosmetic products, rejected the Keystone XL oil pipeline, raised fuel-efficiency standards, and unveiled a clean energy plan.

A new era in environmental regulations is being brought about by the Donald Trump administration along with a major budget decrease for environmental protections. President Trump removed the Obama-era polices aimed at climate changes. Additionally, the Trump administration ended the NASA Carbon Monitoring System, removed the USA from the Paris Agreement, and is reviewing regulations that protect hundreds of threatened wildlife species, to name a few changes the Trump administration is making in regards to environmental protection.

Rural vs. Urban: the New Natural Resource Management Paradigm

According to Brunson et al. (1997), researchers have found little rural-urban variation in attitudes toward environmental policy. Even though urban dwellers impose their values on the resource-based livelihoods of the rural people which in turn cause a greater need for the extracting of resources by the rural people the alliance with the DRMP if their family derives income from the timber production sector or if they believe the policy will economic or culturally sustain their communities.

Conversely, support for NRMP comes from rural and urban non-dependents subgroups (Brunson et al., 1997). Whereas, dependents on the timber industry are less supportive of NRMP, suggesting rural nondependent support slightly less than urban non-dependents suggesting that community dependency is not a strong influence. Personal dependency influences the support for a resource management paradigm more so than urban or rural residency. Community dependency is less influential over support for a resource management paradigm with rural non-dependents only slightly less.

Environmental Hypocrites

According to the research Steel (2009) presents, there is little success in predicting the adoption of environmentally sound behaviors despite wide spread public concern for the environment. Moreover, Steel (2009) finds "significant relationship between political participation and environmental protective behavior" (Findings section, para. 2). Concluding that liberals have more commitment to environmental protection and are more likely to engage in environmentally protective behaviors than their conservative counterparts. Education plays a role in citizen participation in environmental activities with more highly educated respondents reporting the highest levels of political participation in environmental issues (Steel, 2009).

It is unlikely that all citizens in society can change their habits and behavior toward environmental protection. As a society, we have decades of learned behavior that is now being proven to be harmful to the environment; therefore, it seems unfair to categorize a person as an environmental hypocrite because it is essentially impossible for citizens of this society to practice all of the environmentally-friendly options all of the time.

References

Brunson, M., Shindler B., Steel, B. S. (1997). Consensus and Dissension among Rural and Urban Publics Concerning Forest Management in the Pacific Northwest. *Public Lands Management in the West: Citizens Interest Groups and Values.* Santa Barbara, CA: Praeger.

Steel, B. S., Brunson, M., Smith, C. (1993). An Introduction to Social Assessment Techniques for Natural Resource Managers. *Consortium for the Social Values of Natural Resources.* PNW Station, United States Forest Service.

Steel, B. S. & Lovrich, N.P. (1997). An Introduction to Natural Resource Policy and Public Lands: Changing Paradigms and Values. *Public Lands Management in the West.* Santa Barbara, CA: Praeger.

Steel, B. S. (2009). An Introduction to Natural Resource Policy and the Environment: Changing Paradigms and Values. *Public Lands Management in the West.* Santa Barbara, CA: Praeger.

Vig, N. J. & Kraft, M. E. (2010). *Environmental Policy: New Directions for the Twenty-First Century*. Washington D.C.: CQ Press.

Social and Cultural Influence on Environmental Policy: Environmental Policies and Problems Nepal, Korea, the United States and Australia

Environmental Policies and Problems of Nepal

Environmental problems in Nepal including the Himalaya mountain and hill regions, the Terai and Kathmandu Valley consist of overpopulation, urbanization and sprawl, deforestation, soil erosion desertification, land, water and air pollution (Steel & Sushil, 1999). These environmental problems occur regardless of the 1991 Constitution of Nepal that mandates environmental protection and conservation as priorities of the nation. The lack of environmental policies is not due to the lack of requirement by the constitutional, rather from the need to supply the people with basic life necessities at any cost, including degradation of the environment.

The environmental problems of the mountain and hill regions of Nepal include deforestation, soil erosion and desertification. The heavy deforestation of this region results in the erosion of valuable agricultural topsoil at a rate of twelve tons of soil per acre per year during the monsoon rains with the worst places losing eighty tons causing landslips endangering the villagers of the valley and introducing sediment into the rivers (Steel & Sushil, 1999). The continuation of deforestation is a result of, as Steel & Sushil (1999) note, tourism and industrial carpet making due to the large demand of firewood for drying the carpets and for serving the needs of the tourist (i.e. cooking for them and heating for their lodging).

The Terai and Kathmandu Valley is victim of overpopulation, urbanization with unplanned development, as well as air, water and land pollutions. The growing urban population causes problems such as with solid waste disposal as no formal system exists. Particularly the lack of proper disposal of human and animal waste that collects on the streets eventually washing into the rivers and ultimately contaminating the drinking water. According to Steel and Sushil (1999), ninety percent of all piped drinking water contains disease causing microbes and hazardous chemicals. Increased air pollution from factories, motor vehicles and leaded oils affects the residence of Nepal's lower valleys so much so that Kathmandu is now one of the world's most polluted cities where residence now wear face masks to protect themselves (Steel & Sushil, 1999). The continuing pollution along with the increase in the number of diseases namely typhoid, cholera, diarrhea, hepatitis, bronchial infection and skin allergies have economic consequences such as reduced foreign investment, reduced tourism, the need to import drinking water and a weakening agriculture production (Steel & Sushil, 1999).

Socioeconomic and political factors influence the environmental policy process in Nepal through the hands of the elite groups as opposed to the poor populous. The elite community of Nepal consists of Hindus and Buddhist as well as a group of various democratic parties. The Hindus and Buddhist are leftist as to where the democratic groups are rightist. There are several Hindu castes represented (excluding the untouchable castes) in the elite of parliament; thereby the views of the various castes come into play when creating environmental policy. Steel and Sushil (1999) indicates these views toward the environment are pessimistic resulting in incidence of overuse of land, of neglect of maintenance, overgrazing and deforestation. The Buddhist worldview of harmony with nature and animals represents itself in the environmental policy process.

Environmental Policies and Problems of Korean

Korean experienced a fast change to industrialization beginning in the early 1960s that led to severe environmental problems because of the lack of environmental regulations and the drive to reach industrialization through what Park and Kim (2000) describe as the rapid expansion of heavy industry. Economic growth in Korea occurred without regard to the environment or the health of the populous. The preoccupation of developing into an industrialized nation allowed for the dumping of toxic chemicals and waste into the rivers, railway construction through sensitive habitats, urban growth and sprawl without regulation or control, to name a few. The results of the preoccupation of developing into an industrialized nation include municipal and industrial waste clogs the rivers, streams, and lakes, untreated sewage, diminishing wildlife and wildness areas and unplanned metropolitan sprawl in unplanned (Park and Kim, 2000). The Korean government is a parliamentary government similar to that in Britain, with Minister to oversee various aspects of governmental affairs, such as Minister of Environment and the Minister of Construction, and to create policy that parliament may enact. The president is also able to create policy that parliament may enact. The ability of both ministers and the president to create policy sometimes inters with the creation of policy, particularly environmental policy due to the overreaching hand of many ministers and the president as stakeholders in the outcome of the policy. Park and Kim (2000) describe the problem to be those agencies which require other agencies' coordination and agreement in its policy area have weaker political power since they must overcome these other agencies' veto power and pass multiple clearance points for their operation.

Environmental Policies and Problems of Russian Far East

Russian Far East, as part of a post-communist country, has many environmental problems, such as deforestation, extensive mining damage and pollution, water pollution, soil erosion and the lack of subsistence for indigenous populous. The natural resources and the environment of the Russian Far East under the communist rule experienced the most brutal treatment to present. Forests dissemination occurred in order to reach the mineral underneath. Railroad construction occurred without regard to habitat preservation. The communist party built dams without regard in stride toward industrialization. Throughout the push to industrialize and the degradation of the environment, the communist party permanently effected the indigenous populous, not to mention the country's natural resources. The indigenous communities of the Russian Far East under the communist party, lived, under force, collectively with other tribes and usually out of their native habitats. This caused a stress on their subsistence way of life, including their hunting tradition and their custom dwelling construction. Often the indigenous communities in order to survive sought employment at the newly constructed factories and forced to purchase food and life supplies from the communist stores instead of growing their own food. In the post-communist Russian Far East, the indigenous populous now experience a lack of employment, as well as environmental damage that prevents their original sustenance lifestyle not to mention the new presence of foreign investors who purchase Russia's natural resources. The presence of foreign investors and corporations increased the rate of use for the natural resources of the Russian Far East furthering polluting the waters, land, soil and air, overharvesting fishing, clear cutting forests, constructing factories and railroads without regard to sensitive habitats; not to mention the increased strain on a struggling indigenous community.

Environmental Policies and Problems of United States

Environmental policies of the United States receive influence from various stakeholders in the outcome of the implementation of the policy. Unlike the previous mentioned nations, the United States is a republic in which the citizen choice the policymakers through a democratic election. Because the citizen of the United States are directly involved with the selection of the policymakers, they (the citizens) feel a connection with the policies and the outcome of such polices making them (the citizens) active members of the environmental policy process. It is easy for a citizen of the United States to express their concern over the misuse of the environment or in protest of a policy; this is not a privilege easily enjoyed by the entire world's population. Environmental groups share the privilege of expressing dislike for environmental policy and are able to work with the government to change an existing or to implement a new policy. Industrial interest groups also have a stake in the outcome of environment policies and, like the environmental groups, work with the government to implement or prevent the implementation of environmental policy. The United States governmental form is so that all interest groups have the same opportunity to express their concerns and desires over the future of environmental policy.

There are similarities in the environmental problems afflicting the United States and those affecting Nepal, Korea and the Russian Far East, among these are water, air and land pollution, deforestation, and urban growth and sprawl. Unlike, Nepal, Korea and the Russian Far East, the United States, because of its political system, allows for more involvement from the private sector, which ensures the government hears the voice of the people.

Environmental Policies and Problems of Australia

Australia's environmental policy process and the problems associated with the lack of consensus on the implementation of such policies because of differing views of amongst the political parties' results in the lack of action. Strahan (1999) indicates the Australian Labor Party promotes a lean government and views taxes as burdensome, punishing disincentives. Moreover, according to Strahan

(1999), the Australian Labor Party believes the government should not compete with private sector, noting their political philosophies is in-line with the conservative end of the Republican Party spectrum in the United States.

On the other side of the political spectrum in Australia is the Liberal Party of Australia is a democratic socialist party that believes in the socialization of industry, production, distribution and exchange as a means to eliminate exploitation and wants to redistribute power and wealth as a means of maintaining a non-monopolistic private sector (Strahan, 1999). In Addition, the three established goals of the Liberal Party of Australia, as Strahan (1999) indicates, regarding natural resources include, social ownership, sustainable utilization, and recognition of the need to sustainably develop.

As evident by the differing views of the political parties in control of the environmental policy process, the utilization of natural resources and the destruction of the environment may not receive immediate action. The Australian Labor Party favors corporate utilization of resources that are the distributed to the people of Australian and exported worldwide; this also coincides with the behavior of the democratic rightist parties of Korea. Conversely, the Liberal Party of Australia favors sustainable utilization of Australia's natural resources for the greater good of the populous instead of for private profits.

Cultural/Religious Traditions and the Potential Impact on Environmental Policies

Hindus and Buddhist traditions effect on environmental policy

The Hindu religion, as well as the Buddhist religion, believes in rebirth and karma. The Hindus believe each human and animal composition includes, as per Coward (1995) a combination of spiritual and non-spiritual elements, as well as the most extreme case of Jainas whom include plants, rocks, air and water. This view of the spirit and non-spirit existing in all aspects of nature in addition to the idea of rebirth, the Hindus view of nature is to take care of it in order to preserve their home for future generations.

The Buddhist view of the karma cycle does not include animals, plants, rocks, water or air, as does the Jainas of the Hindu religion (Coward, 1995). The Buddhist shares the same oneness with nature observed by the Hindus. More preciously the Buddha taught that to "exist in any sense at all means to exist in dependence on the other, which is infinite in number and that nothing exists truly in and of itself, but requires everything to be what it is" (Coward, 1995, p. 9). This view of dependence on nature influences the Buddhist participants in the environmental policy process.

Western religions varying view on nature and the effect on environmental policy

Christianity, Judaism and Islam, the three major western religions, share an anthropocentric view on humankind's relationship with nature and the view that humans are to be fruitful and multiply and fill the earth and subdue it, yet there are very distinct differences in the interpretations and implementations of these shared views (Coward, 1995). The Christian view of humanity's role with nature is one of dominance where humans are to rule over nature and animals which resulted in the cultures' being responsible for the ecologically unsustainable patterns of development.

Islam, unlike Christianity and Judaism, views nature (and the Quar'an) as God's self-revelation believing the cosmos is the manifest of God's compassionate breathe through its regularity and beauty (Coward, 1995). Following the belief that nature displays God's potentialities and attributes. One would suspect that with a strong connection between god and nature that the Islamic influence of environmental policy would be to preserve nature and treat nature with respect as they would god.

Judaism, like Christianity, views their role with nature as stewards of the land. Jews in keeping the *Shabbat* enforces the idea of stewardship of nature. "We join together, in harmony, with the sky, the land, the seas, the animals of the waters and of the lands, the planets, stars and galaxies in recognizing Gods as the Life and Strength of the universe" (Coward, 1995, p. 87) on the Shabbat the people come together for celebration of recognizing their part in creation.

Conclusion

Environmental problems are similar throughout the world, including overpopulation, deforestation, water, air and land pollution, toxic waste disposal and all the after effects of such activities. The difference comes from how the various governments of different countries handle environmental problems and the creation of the policies to tackle such problems. Although Nepal, Korea, the Russian Far East, the United States and Australia share similar environmental problems, the structure of their governments directs the solution to the problem and thus the outcome. Each country has various stakeholders who also have an outcome of environmental policy process; and depending on the country, the ability of the stakeholders to voice their opinions varies.

References

Coward, H. (1995). *Population, Consumption, and the Environment.* New York, NY: University of New York Press.

Park, C. & Kim, S. (2000). Environmental Policy in South Korea. *Handbook for Global Environmental Policy and Administration.* New York, NY: Marcel Dekker.

Steel, B. S. & Sushil P. (1999). Political Elite Commitment to the Environment in Nepal. *Global Environmental Policy and Administration.* New York, NY: Marcel Dekker Publishers.

Strahan, E. (1999). Comparative Environmental Policy: Australia and the U.S. *The Environment and Socioeconomic Issues: A Common Thread.* Cambridge, MA: MIT Press.

Forces That Drive Environmental Policies

Environmental policies exist to direct the course of behavior of citizens, municipalities and industry nationally and internationally with the intention of protecting the health of the environmental, the populous and natural resources. Depending on the country's level of economic development, namely post-industrial, developing and post-communist nations, the forces that drive environmental policy vary. Such forces include the type of economic structure (e.g. capitalism); governmental structure; and the involvement of non-institutional interest groups and stakeholders.

Economic structure

Post-industrial countries, such as the United States and Australia, generally, have a capitalist economy. The United States and Australia are both market-based economies. Australia exports agricultural products, minerals, metals and fossil fuels deriving most of their income from raw products opposed to manufactured goods as in the United States (Strahan, 1999). This makes the Australian economy vulnerable to shifts in the commodity market a trait typically associated with undeveloped not developed countries. Environmental policies affect the capitalist economy through the establishment of regulations regarding air emissions, solid and effluent toxic waste disposal, and material safety, to name a few.

Developing countries, many of which are in African, have no economic structure – most are purely sustenance societies that struggle daily to provide for their family. Starvation and dieses infiltrate the population in developing countries furthering inhibiting economic growth. Most developing countries rely on foreign aid, such as the United Nations, to feed their citizens. India and China are developing countries with established economic structures; however, their quest to industrialize often comes with environmentally harmful industrial practices.

Post-communist countries, such as Russia, for this essay specifically the Russian Far East (RFE), are transitioning from a communist nation to one a capitalist economy resulting in an emerging entrepreneurial class that is eager to exploit newly available natural resources with little regard for the environmental impact (Rineer-Garber, n.d.). During the communist rule of Russia, the communist officials rounded-up the various groups of indigenous people living in the RFE putting them into combined communities, usually away from their traditional means of sustenance. In a post-communist RFE, private entrepreneurs and foreign corporations are extracting natural resources in a matter that is not supportive of the indigenous populations' way of life as well as destructive to the environment. However, with a struggling economy, the RFE groups accept money in return for their way of life.

The post-communist nations that transitioned into capitalist economies experience conflicting views regarding the use of natural resources, thus affecting the formation of environmental policy. Indigenous groups conflict with private industry over the use of RFE natural resources. Private industry views environmental policy as either helpful for their ends or as a hindrance to their ventures. Compare this view with that of the indigenous groups that rely on natural resources as a means of survival and environmental policies that deregulate industrial practices often harm the source of the indigenous population's survival. According to Strahan (1999), "this is just another blow to the ethnic minorities of the Russian far north who have seen their traditional lifestyles destroyed over the past decades by forced resettlement and ill-considered industrialization" (p. 26).

Governmental structure

Post-industrial nations such as the United States and Australia have democratic governments, although each is slightly different. The United States has a presidential democracy in which the people elect the president and the members of congress; separate executive, legislative and judicial branches and where the president is both the head of the government and head of state (Strahan, 1999).

The separation of the governmental branches in the United States allows for a checks and balance system in an attempt to prevent corruption and to ensure a valuable government to the people. The separation of governmental branches in the United States, at times, seems to slow the progress of needed environmental policies because of the numerous procedures involved with the separation of powers.

Compare this with the Australian parliamentary democracy in which the legislative branch the Parliament-selects the Prime Minister who shares executive authority with the chief of state of state-currently United Kingdom's Queen Elizabeth II represented by her appointed Governor General (Strahan, 1999.). Other governmental participates include the different political groups that governmental officials represent. Australia has a coalition government in which all the smaller political groups joined forces with one of the two major political parties, namely the Australian Labor Party and the Liberal Party of Australia (Strahan, 1999).

The Australian Labor Party promotes a lean government and views taxes as burdensome, punishing disincentives. The Australian Labor Party is of the opinion the government should not compete with private sector noting their political philosophies are in-line with the conservative end of the United States Republican Party spectrum (Strahan, 1999). Conversely, the Liberal Party of Australia is a democratic socialist party that believes in the socialization of industry, production, distribution and exchange as a means to eliminate exploitation (Strahan, 1999). Additionally, the Liberal Party of Australia wants to redistribute power and wealth as a means of maintaining a non-monopolistic private sector. Moreover, the three established goals of the Liberal Party of Australia regarding natural resources include, social ownership, sustainable utilization, and recognition of the need to sustainably develop.

Each of Australia's political parties influence environmental polices based on their groups belief system. There is a clear difference between the Australian Labor Party and the Liberal Party of Australia view the government's role in private business and natural resource ownership. The Australian Labor Party favors deregulation as a means of promoting private sector industry; conversely, the Liberal Party of Australia favors socialism of industry and the ownership of resources. This one difference greatly affects the outcome of environmental policy.

Developing nations range in governmental structure from democracies like India's to military dictatorship like the one recently voted out of office in Niger. The United Nations Development Programme reports the people of Niger recently went to the voting polls to elect a new president and parliament, which completes transition from military to civilian rule (UNDP, 2011). This is a major step for developing nations in terms of both environmental policy and government development. Without an established civilian government, as in the case of Niger, the United Nations, provides for the citizen services that a develop food donation, building materials, education services, infrastructure construction and medical care all of which typically comes from the government. Environmental care, I imagine, is far from the minds of developing countries. In my opinion, the ruling bodies of these countries concerns are more on the growth of their economy than on the health of the environment. Developing countries have concerns that focus on current situation rather than, in my opinion, my sustainable practices of development.

Post-communist countries experienced a transition from a communist rule to a democratic government affecting the responsibilities of the local governments. According to Cathleen Rineer-Garber (n.d.), "it was predicated that decentralization [of the communist party in Moscow] would permit local authorities to respond to immediate environmental concerns, while the liberation of the media and formation of environmental organizations would force accountability on the government and industry" (p. 1). However, this was not the case; the exploitation of resources is occurring while little regard to the environmental impact all as a means of building the transitioning government and economy because of the new autonomous and semi-autonomous regions that gained political power and now display a yearning to maintain control of their natural resources (Rineer-Garber, n.d.).

Non- institutional interest groups and stakeholders

Post-industrial countries have various non-institutional interest groups and stakeholders (e.g. environmental groups, corporations, citizens) that influence environmental public policy. In the United States, the idea that science should assist in the resolution of environmental and natural resource issues by suppling scientific information to more successfully inform policy and management (Steel et al., 2008). All non-institutional interest groups can use scientific information to promote their goals. For instance, environmental groups as a means of protecting habitat can influence environmental policy by referring to scientific data that supports their claims. The same can be true for industry, industrial leaders will use scientific data that supports their industrial purpose, and for instance, the lack of damage the company will inflict on the environment.

Developing Nations

Post-communist countries now embrace some form of democratic reform and adopted some type of market-based economic practices which allows them to engage in science-based environmental policy, which, according to Steel et al. (2008) is now pronounced as important in many post-communist regimes. Although, environmental policies form based on scientific data, the policy is not always enforced and in a post-communist country, enforcement is more difficult than in a post-industrial nation. Moreover, the formation of joint ventures between Russian and other countries (government and private industry) have further increased the rate of destruction while little profit is enjoyed by local people. Additionally, the Russian populous, according to Strahan (1999), hold the belief that the nations' resources, including the forest, belong to all Russians and some favored elite. Regardless of how the Russian population feels about the industrialization and private industry, they willingly work for the companies, as it is becoming their only means of survival.

Simultaneous solutions

Developmental problems (such as the lack of electricity) and environmental problems (such as population from coal plants) have the opportunity to share a solution. For instance, if a developing country needs more electricity the solution, instead of a coal burning plant is solar power, wind turbines or hydroelectric power. Controversy exists over hydroelectric dams and there damage to fish population; while this is true in certain situation, it is not always the case and deserves considering when discussing an increase in the production of electricity. Water shortages and water purity are problems in some countries; a shared solution is developing a water waste treatment facility that in turns waters (and feeds) the vegetable gardens. Both of the simultaneous solutions previously discussed will work in post-industrial, developing and post-communist nations. One can be optimistic humanity will adapt to the necessary changes required in order to ensure the quality of life for future generations. With that said, one may be concerned post-industrial countries (namely the United States) will not change the ways of deregulating industry instead of holding industry accountable for its contribution to a poisoned environment. One can be less optimistic about developing nations' willingness to develop clean and green methods of economic growth instead of following the polluting pattern followed by post-industrial nations.

References

Steel, B. S., Warner, R. & Johnson, A. (2008). Environmental NGOs and Science Policy: A Comparative Analysis of Bulgaria and the U.S., *Journal of Environmental Systems, 31*, 141-157. Thousand Oaks, CA: Sage Publications

Strahan, E. (1999). Comparative Environmental Policy: Australia and the U.S. *The Environment and Socioeconomic Issues: A Common Thread.* Cambridge, MA: MIT Press.

Rineer-Garber, C. (2003). Transition to a Free Market Economy in the Russian Far East: The Environmental and Social Consequences. *Handbook of Global Environmental policy and administration.* Boca Raton, FL: CRC Press.

United Nations Development Programme. (2011). UNDP supports Niger elections. Retrieved from http://www.undp.org/content/undp/en/home/presscenter/articles/2011/01/28/undp-supports-niger-elections.html

Driving Forces of Environmental Politics and Policy

There are several driving forces of environmental politics and policy including population growth and urbanization, economic expansion, technological changes, value change and globalization. Differing views on the benefits and/or hindrances of the various driving forces of environmental politics play a significant role in the outcome of the policy. Leaders, citizens and industry in developing, post-communist, and post-industrial nations have varying opinions on environmental politics and different stakes in the outcome of natural resource policy.

Population Growth

There is startling data about population growth, such as the world's population reaching 7 billion people in October 2011 and increases at a rate of 90 million people a year (Roser and Ortiz-Ospina, 2017; WorldOMeters, 2018). Along with the population growth is the rapid growth of urban areas with larger portions of the population moving out of the farmlands and into the city. Currently, more than half: of the world's population live in urban areas notably twenty of the world's largest urban areas have populations greater than fifteen million (World Bank Group, 2018; Bate, 2016). The growing population increases the use of natural resources affecting each person differently depending on their soci-economic background. What environmental groups might view as preserving forests for endangered species' habitat as a positive action the indigenous people might view the same action as a negative effect on their personnel substance for which the forest provides. The same is true for water rights issues, air quality problems and water purity concerns not to mention the increase consumption of natural resources and waste production all of which stem from rapid urbanization resulting from population growth. Additionally, urbanization has a profound effect on ecosystems and human health worldwide while causing traffic congestion and serious air quality problems in the post-industrial nations.

Economic Development

A country's economic structure, whether dominated by labor-intensive employment as is the case in less developed nations or dominated by the service sector as in post-industrial nations affecting the function of society, interpersonal relationships, technological achievements, and the structure and operations of its political system. As does any change or shift in their economic system. The agrarian lifestyle found in most developing nations maintains living conditions considered at poverty level and subsistence level for the majority of its population, so when a shift occurs economically and industrialization begins large portions of the agrarian populations move into urban areas in pursuit of employment. Industrialization occurs in phases with the first phase including a move away from environmentally sustainable transportation such as sail-boats and equine-energy toward steamboats and locomotives that run on coal energy allowing for an increase in productivity distribution and consumption of goods and services. The shift from coal to oil occurs during the second phase of industrialization with an increase in electrical use, technological advancement and mass consumerism coinciding with the use of the automobile, increasing oil dependency while reducing urban air quality. Post-industrial countries maintain a higher standard of living resulting from a larger population of knowledge intensive service sector jobs, which can be accredited to the enormous technological innovations and mechanization in agriculture and manufacturing sectors.

Technological Change

The role of technology as it affects environmental politics and policy differs depending on point of view. The followers of the dominant social paradigm (DSP) maintain a cornucopian viewpoint in which free market forces and human ingenuity will lead to new technologies that will meet the demands of a growing world population. It is worth noting environmental scientist and policymakers benefit from technological achievements providing tools allowing for more accurate assessments of environmental conditions. Conversely, the followers of the new environmental paradigm (NEP) are more in line with the Neo-Malthusians that believe technological advancements will not keep up with the increasing resource needs of an ever growing world population instead they believe technology is the source of pollution, climate change and biological/nuclear warfare.

Value Change

Industrialization, as mentioned, coincides with urbanization ultimately shifting the ideals associated with an agrarian society that maintains a collective focus toward an urban society whose population focuses more on the individualism and materialism. Post-industrial countries values are shifting toward the NEP resulting from individual value shifts toward obtainment of a higher quality of life rather than subsistence needs. Data suggests environmental awareness is also prevalent in developing nations with protecting the environment having a higher priority than economic growth (Thai, et al., 2007).

Globalization

There are multiple dimensions of globalization including environmental, cultural, political, trade, finance, labor and technology. There is a general understanding that globalization refers to the exchange of capital, services, and goods with increased levels of economic activity integration (BERA, 2012). Proponents of globalization have principles harmonious with those of the DSP and adhere to five major claims including: (1) only in a democratic society can free markets exist because of the protection of individual freedom; (2) technological innovations drive the spread of irreversible market forces; (3) globalization is guided by market forces; (4) the expansion and liberalization of world trade benefit everyone; (5) democracy will expand throughout the world because of globalization (Steel, et al. n.d.). Additionally, economists promote globalization and free trade referring to the monetary gains obtained by wealthier nations and their ability to preserve the natural environment.

On the other hand, opponents of globalization fear the homogenization of international cultures and lifestyles where a small number of multinational corporations retain the concentration of power ultimately rendering detrimental consequences. Opponents of globalization are critical of neoliberalism, which they believe deepens the social divisions between rich and poor (Ehrke, 2001). Environmentalist often oppose globalization reasoning that free trade accelerates the process of global environmental degradation in the pursuit of national wealth (Lofdahl, 2002).

The impact of the forces behind environmental politics and policy are without consensus amongst academics. Two major paradigms influence the discussion amongst academics, namely the dominant social paradigm (DSP) and the new environmental paradigm (NEP) (Steel et al., 2004). The DSP represents a belief society accepts as true that it is seldom questioned with the ability to direct public opinion regarding public policy. Additionally, the DSP influences the decisions made by government officials whose current focus emphasizes growing of the economy, expanding trade amongst countries, and rely on scientific and technological advancements. The anthropocentric view in which the environmental and natural resources are for the benefit of humankind held by the DSP contrasts harshly with the biocentric belief of the NEP. The NEP are of the opinion that humans co-exist with the environment and focus on the systematic protection and preservation of ecosystems. Additionally, the DSP and the NEP disagree over the role that science and technology influence environmental politics and policy. Proponents of DSP argue technological advancements and scientific breakthrough will have a solution to all problems while supporters of the NEP counter that scientific and technological progress come with a great cost to the environment, for instance the agricultural industry's dependency on the chemical fertilizers and pesticides.

Similarly, discord exists in the scientific community regarding the forces driving environmental politics and policy. Science in and of itself is complex and full of uncertainties often leading to differing consensus amongst scientific groups. An example of the conflict amongst the scientific community was obvious in the debate over water rights in the Klamath Basin. There was conflicting scientific reports on the status of endangered species and the amount of water necessary to sustain the fish population with the U. S. Fish and Wildlife Service along with the National Marine Fisheries Service disagreeing with the National Academy of Science (Steel et al., 2004).

Global environmental politics and policy involve various institutional and non-institutional actors on a local, national and international level from perceiving a problem, developing an issue, getting inclusion in the government agenda and lastly securing government action.

Institutional Actors

Governments and associated agencies, legislatures, executive departments and judicial systems are institutional actors involved in the process of environmental policy. Additionally, various international organizations play a role in leading, supporting, or blocking issues of internationally significant environmental concerns, examples of such organizations include the International Whaling commission (IWC), United Nations Environmental Programme (UNEP), the United Nations Development Programme (UNDP), and the European Environmental Agency (EEA).

Non-institutional Actors

Non-institutional Actors include, non-governmental organizations (NGO), environmental groups, ecological resistance movements, scientific community, the media, political parties, and corporations. Each non-institutional actor has a different involvement in and stake in the outcome of environmental policy. Non-governmental agencies are non-profit groups that operate with autonomy focusing on humanitarian or cooperative objectives instead of commercial intentions and include large-scale membership groups, community-based groups, international organizations, think tanks and activists' organizations. Ecological resistance movements emerge from grassroots movements lead by peasants, indigenous people, economically marginalized populations and nature groups challenging the anthropocentric view of natural resource management noting it as the cause of environmental problems often participating in civil disobedience, boycotts and / or the destruction of property. The ecological resistance movement in post-industrial societies occur as environmental justice movements targeting the injustice imposed on people of color and working class whites who experience social inequality regarding environmental degradation, resource depletion, pollution and environmental hazards. Scientist as participants in the process of environmental policy provide their expertise toward understanding environmental issues and problems adding an important component to the efficacy of environmental public policy. Moreover, the role of science and scientist in the environmental policy process is debatable with several opposing views on the relevance science brings to the influence and creation of environmental policy.

Science and scientists provide information to policy makers based on scientific theories and experimental information. Each branch of science provides a focused view of the various environmental problems; thereby allowing the policy makers the opportunity to pick and choose the data that best fits their purposes. Currently, there is not a consensus between scientist concerning environmental problems. One side of the argument contents scientific discoveries foster new technologies that, in the past, have provided solutions to environmental problems. The other side of the argument maintains science cannot keep up with the demand the rapidly growing population is putting on the available natural resources. Additionally, there is no consensus amongst policy makers that science plays an important role in the environmental policy process claiming that science may in fact get in the way of progress. If policy makers are able to pick and choose the science that fits the means for their ends; then science may impede progress.

The environmental policy process is complex because of the numerous participants and the dramatically varying viewpoints on how to manage natural resources that ultimately affects the condition of the environment. Many believe the current DSP is leading to irreversible damage to our natural resources and environmental and embrace the NEP that focusing on a biocentric approach to resource management. With the assistance of various institutional and non-institutional participants policy makers deliberate on environmental concerns facing our society making eventually making consensus on the correct course of action and finally implemental a plan that will, hopefully, sustain the natural resources for current and future use.

References

Bate, O. (2016). A countdown of the 20 largest urban areas on the planet, according to the 2015 edition of Demographia's World Urban Areas index. *Top 20 megacities by population.* Retrieved from https://www.allianz.com/en/about_us/open-knowledge/topics/demography/articles/150316-top-20-megacities-by-population.html/#!m8ef2a55a-d6ac-41a7-a675-40521793b00f

Business and Economics Research Advisor. (2012). Defining Globalization. *A Series of Guides to Business and Economics Topics, Globalization, 1.*

Ehrke, M. (2001). What do the opponents of Globalization want? *Friedrich Ebert Stiftung, Department for Development Policy - Dialogue on Globalization.*

Lofdahl, C. (2002). *Environmental Impacts of Globalization and Trade.* London: MIT Press.

Roser, M., Ortiz-Ospina, E. (2017). World Population Growth. *Our World in Data.* Retrieved from https://ourworldindata.org/world-population-growth#estimates-of-population-in-recent-history-and-projections

Steel, B., List, P., Lach, D., Shindler, B. (2004). The role of scientist in the environmental policy process: a case study from the American west. *Environmental Science & Policy, 7,* 1-13.

Thai, K. V., Rahm, D., Coggburn, J. (2007). *Handbook of Globalization and the Environment.* Boca Raton, FL: Taylor & Francis Group.

World Bank Group. (2018). Urban population growth. Retrieved from https://data.worldbank.org/indicator/SP.URB.GROW?locations=BI&year_high_desc=true

World O Meter. (2018). *World Population.* Retrieved from http://www.worldometers.info/

Environmental Policy Process Participants

Environmental policy issues in the United States both at the national and local levels receive influence from various participants including nongovernmental interest groups and governmental agencies. Government agencies are responsible for generating, regulating and enforcing environmental policy. On the hand, nongovernmental agencies work to influence the policy process in a manner that will facilitate their goal. All participants involved with the environmental policy issues have a stake in the outcome of the policy whether it is to preserve land, regulate logging or monitor carbon emissions, the stakeholders involved want the satisfy their own objectives. The resources available for the participants to use, various in both type and quantity, influence the outcome of their purpose.

Non-governmental Interest Groups

Non-governmental organizations (NGO), such as environmental interest groups and industrial organizations, utilize human, financial and organizational resources. NGOs in the United

States assemble like-minded citizens for collective action and represent the citizen groups' interests in government (Vig and Kraft, 2010). These interest groups use diverse strategies to influence public policy, such as lobby elected officials and bureaucrats, organizing grassroots movements to mobilize public opinion, build coalitions and make monetary contributions into the political system (Steel, n.d.). Interest groups chose which strategies to use based on the accessibility of the resource, the quantity of resource available and the overall effectiveness of the strategy.

There are two fundamental difference between the membership bases of NGOs. One membership base is comprised of individual citizens the other is an aggregations of representatives of government, institutions, and businesses. Although, both groups use the same types resources and strategies, their origins and purposes are drastically different. NGOs that are part of the environmental movement, typically, are the product of urbanites, have post-industrial values, and support the New Environmental Paradigm (Steel, n.d.). Environmental NGOs, such as the Natural Resource Defense Council (NRDC), the Environmental Defense Fund and the Wilderness

Watch, came into being after Congress passed NEPA. Moreover, Environmental NGOs, such as the Sierra Club and the National Wildlife Federation, underscore the importance of resource preservation as to provide for aesthetics, wildlife habitat, and the wilderness experience (Davis, 2001). In contrast, support for industry-related interests, typically, come from rural and sparsely populated nonmetropolitan areas dependent on natural resource extraction (Steel, n.d.).

Public interest groups' purpose is communication and educating the public. Environmental groups, such as the Sierra Club, National Audubon and National Wildlife Federation and grassroots organization are designated as public interest groups, citizen groups and social movements. Public interest groups target specific groups of citizens that support and reinforce their agenda. There are categorical divisions amongst public interest groups including environmental protection and conservation; intensive recreation, and passive recreation. More specifically, individual membership groups' interested in the preservation or conservation of federal forestlands are considered an environmental protection and conservation public interest group. The membership groups representing fishing, hunting, snowmobiling, etc. are interested in intensive use of federal lands and are considered intensive recreation public interest groups. Lastly, passive recreation public interest groups represent activities such as hiking, cross-country skiing, mountain climbing and wildlife watching essentially, those interested in more passive and potentially less environmentally damaging activities.

Public interest groups are substantially influential despite their financial resources due to their grassroots organization and their value of congruency with the public engaging in elite challenging political activities, such as demonstration and boycotts. Elite challenging activism embodies a form of political action that addresses specific policy goals in which the public opposes the existing political agenda and seeks to impose constraints on policy makers.

Public interest groups play an important role in the federal policy debate having grown in number and size over the years despite their narrow focus toward educated, typically white, middle-class Americans, which contribute the bulk of their political, ideological and financial support (Vig et al., 2010). Growth in membership can be contributed to a perceived notation of serious threats to the environment and/or a desire to be out in nature with others sharing similar interest. Moreover, public interest groups rely on hefty memberships and numerous generous financial contributors in order to financial their organizations.

Environmental organizations and grassroots groups utilize the benefits of their large memberships when determining their strategies. Public interest groups rely on letter writing, public demonstrations and volunteers as strategies to accomplish their purpose. These elite-challenging tactics along with political organizing and media campaigns give strength to environmental groups. Environmental interest groups influence derives primarily from their capacity for mobilizing human resources and tend to be values-based and scientifically-oriented information development and dissemination.

Environmental interest groups can exist in, historically, unlikely sectors of society. For instance, the Evangelical Climate Initiative is a group of over eighty evangelical Christian leaders whose purpose is to fight global warming supporting market-based incentives to reduce greenhouse gases (Vig & Kraft, 2010). The

Evangelical Climate Initiative seeks to convince fellow Christians that combating global warming is a moral question sanctioned and that all Christians are to be good Stewards of the Earth (Vig & Kraft, 2010).

Industrial organizations

Industrial organizations, for instance the timber industry and any industry-related support groups, rely on their substantial financial resources to influence public policy giving businesses a privileged position in the American policy process. Groups aligned with geographically defined and economically based constituencies, such as wheat growers and coal miners, are able to speak to aggregate and mobilize their adherents in a more sustained and targeted fashion (Vig and Kraft, 2010. Industrial organizations generally fit into two categories: industry-support groups and industry. Membership groups representing natural resource extraction interests on federal forestland that are not directly involved in commodity production instead represent worker, community and industry interest involved in commodity production are considered industry-support groups. Whereas industry industrial organizations are non-member commercial organizations interested in natural resource extraction from federal forest lands.

Industrial groups' network creating a common front when working in opposition to environmental movements giving them the congruency they uses to take advantage of the slow legislative progress. Industrial groups represent business and professional interests of the private sector giving them an advantage of being well financed and well-staffed resulting in their use of tactics that work for a small group. Despite their few in numbers, as mentioned, the industry-interested groups possess large budgets with which they use to influencing the election of key decision makers or lobbying such decision makers after the election. Testifying before courts and Congress and appealing to various governmental agencies is most effective for industrial groups where they can use their small numbers to speak to the most important policy decision makers. Moreover, industry and industry-supported groups possess relatively substantial financial power, however, have less support and confidence from the public. Additionally, industry and industry-supported groups tend to focus their attention on more traditional forms of influence, such as lobbying natural resources agencies and elected officials.

Coalitions of interest groups

Coalitions of interest groups membership consists of small groups or businesses rather than individuals. Coalitions of interest groups network together to influence federal policy, such as the Environmental Council of the United States and the National Association of State Energy Officials. Coalitions not only represent environmental and industry groups, they also represent States' interests, for example, the Great Lakes Commission and the Western Governors Association. Not only do networks facilitate information exchange, they foster the diffusion of innovation and pool resources to pursue joint initiatives.

Coalitions of interest groups have impressive political force because of their collective financial resources and lack dependency of membership dues. The support coalitions receive from foundations, wealthy individuals and the government is what allows them to reduce their reliance on individual membership dues. Coalitions also have patrons outside the group that provide both financial resources but also networking resources. In recent years, networks of professionals working in similar capacities have become increasing influential (Vig and Kraft, 2010).

Coalitions of interest groups, in recent history, are forming out of, historically, opposing groups. An example of this is the Apollo Alliance, a professed blue-green coalition that has the recognition of shared goals and common enemies bringing together environmental groups and industrial unions. The goal of Apollo Alliance is to promote a national effort to create more green American manufacturing jobs in clean coal technologies, hybrid automobiles and bus transportation infrastructure, and to form a united effort to promote global fair trade (Vig and Kraft, 2010). The power behind such coalitions may prove to be very powerful since they have such large memberships. Another example is the Blue Green Alliance, which promotes job-creating solutions to combat global warming and represents close to four billion people (Vig and Kraft, 2010).

Governmental Participants

Governmental participants influence environmental policy differently, some create policy with others enforce it. All three branches of government have a designated role in the environmental policy process. The President and Congress create subcommittees to manage the details of the policy and the subcommittees create organizations, such as the Environmental Protection Agency (EPA), to implement policy protocol. The Forest Service is a governmental agency that directly interacts with the public on the enforcement of policy procedures.

President

The President of the United States' role as Chief Executive has historically been the most important role when analyzing environmental policy. Along with the President's executive duties, such as appointing cabinet members and overseeing the regulatory process, the President has a leading role in enacting environmental legislation and in gathering public support for new environmental policies. Additionally, the President has a significant role over the implementation of environmental protection in the social, political, and economic sectors of the country (Anderson, 2012). Presidential powers include, agenda setting, policy formulation, legitimate policy, policy implementation and assessment and evaluation of existing polices and propose reform. Furthermore, regardless of a President's ambitious environmental agenda pursuits (which is enabled through executive authority) conflict can arise with industry, States, and Congress (Konisky and Woods, 2016).

Congress

Congress passes and enacts laws that establish environmental policy, such as the National Environmental Policy Act (NEPA) in 1969 and the Endangered Species Act (ESA) (Vig & Kraft, 2010; Davis, 2001). Additionally, Congress creates subcommittees, such as the Public Land Law Review Commission (PLLRC), which are responsible for reviewing public land policies and suggesting changes as well as the direction of such changes (Davis, 2001). On an annual basis, Congress appropriates funds to the operation of the Environmental Protection Agency (EPA) as well as for agencies supporting natural resource protection and extraction and energy research and development. Through its budgetary decisions, Congress can significantly affect executive agencies ability to function effectively, such as the EPA, Department of Energy, U.S. Geological Survey, Fish and Wildlife Services, Bureau of Land Management and Forestry Service (Vig & Kraft, 2010). In addition, through Congressional budgetary oversight, Congress can underpin a dominant sub-government by cutting the sub government's budget. Such was the case in the mid-1940s when the head of the Grazing Service (now the BLM) announced a tripling of grazing fees Congress responded by cutting their budget in half (Davis, 2001). Moreover, Congress affects environmental policy through regulatory reform, governing statutes and appropriations, such as riders. This gives Congress influential powers and control over natural resource management often with accompanying criticism (Vig & Kraft, 2010). Government inaction toward important public problems is usually the result of policy gridlock referring to an inability to resolve conflicts in a policymaking.

Nongovernmental Resources

There are various types of resources available to both public interest groups and industry groups. The resources include full-time and part-time staff, volunteers, members, income from membership dues, donations from foundations and money from business (Steel, n.d.). Environmental groups rely more on human resources than do business interests, which have greater economic resources. Of the average environmental groups' annual budget, over fifty-seven percent comes from membership dues or contributions as compared to industry-related membership groups, which derive only thirty-five percent of their annual budget from membership dues (Steel, n.d.). Business revenue as a source of income for industry-related membership groups comprises about thirty percent of their annual budget compared to environmental groups, which receive less than three percent of their annual budget from business revenue. Not all interest groups have the same amount of resources to influence policy progress. Generally speaking, the groups that fare the best financial are those that purchase and conserve land, for instance, Ducks

Unlimited, Nature Conservancy, Conservation Fund, and Conservation International (Vig and Kraft, 2010).

References

Anderson, E. (2012). A Stronger Role for the United States President in Environmental Policy. *Student Theses 2001-2013*. Fordham University. Retrieved from https://fordham.bepress.com/environ_theses/38/

Davis, C. (2001). *Western Public Lands and Environmental Politics*. Boulder, CO: Westview Press.

Konisky, D. and Woods, N. (2016). *Publius: The Journal of Federalism,* 46:3. 366–391. https://doi.org/10.1093/publius/pjw004

Steel, B., Pierce, J., Lovrich, N. (1996). Resources and strategies of interest groups and industry representatives involved in the federal forest policy. *The Social Science Journal, 33:4*. 401-419. Retrieved from https://andrewsforest.oregonstate.edu/sites/default/files/lter/pubs/pdf/pub4727.pdf

Vig, N. J. and Kraft, M.E. (2010). *Environmental Policy: New Directions for the Twenty-First Century*. Washington D.C.: CQ Press.

Markets, Science, Conflict, and Sustainability's Effect on Environmental Policy

Role of Markets in Environmental Policy Process

Economic efficiency is finding the greatest difference between total benefits and total costs (Vig & Kraft, 2010). Application of benefit-cost analysis to the policy progress finding the equilibrium between the marginal benefits and the marginal costs a technique that often draws criticism for neglecting to consider important political and ethical values (Vig & Kraft, 2010). Critics object to placing a value on natural resources. The social problems associated with the principles of market-based environmental policy are externalities, public good and the tragedy of the commons. Externalities are costs not directly associated with the transaction between the buyer and seller, such as the carbon emissions associated with producing electricity.

Two uses of market principles to solve environmental problems are tradable market permit programs and taxes. Policies involving environmental issues, such as reducing air pollution establishing quotas for individuals fishing for trade, regulating waste management and habitat management currently utilizes market-based principles. Oregon House Bill 2186 Metropolitan Planning Organization Greenhouse Gas Emissions Task Force exemplifies Oregon's dedication to reduce greenhouse gas emissions. HS2186 requires the creation of a Metropolitan Planning Organization (MPO) Greenhouse Gas Emissions Task Force to "evaluate alternative land use and transportation scenarios that would meet community growth needs, while reducing greenhouse gas emissions and recommend future legislative action to support such efforts" (ODOT, 2010). Pew Center on global climate change recognizes the usefulness of cost-benefit analysis (CBA) as an important tool in the environmental policy process as well as recognizing the criticism CBA receives for being insufficient as the primary approach to environmental issues (Pew Center, 2010).

Role of Science in the Environmental Policy Process

Science as a factor in the process of creating environmental policy brings with it conflicting proposals, different interpretations and dueling data, which gives fuel to its opponents. The scientific community favors the method of risk-based decision-making. The Environmental Protection Agency (EPA) describes risk-based decision-making as the process that "utilizes risk and exposure assessment methodology to help [the projects] implementing agencies make determinations about the extent and urgency of corrective action and about the scope and intensity of their oversight of corrective action by [the projects] owners and operators" (EPA, 2009).

Risk management is the process of evaluating the policy options concerning controversial issues in a dialogue with all the persons with a direct interest, involvement or investment in the issue (EurActiv, 2005) Analyzing the risks, both entailing and those perceived by the public, policy-makers not only make decisions about the course of action to take regarding the risk but they also need to communicate their decision, oversee its execution and appraise the outcome.

The scientific community and industry favor risk-based policy making indicating risk analysis as the only objective scientific basis that can pilot more rational decision. Advocates proclaim set-backs and restricts of risk analysis are capable of being overcome by means of systematic collection of data and studies with rigorous guiding principles for dependable conduct and presentation of final outcomes. Critics of risk analysis, typically environmentalist, consumer interest groups and other NGOs express concern regarding the oversimplification of issues by policy makers that focus on a problem in unilateral fashion. Concerns over political manipulation in the environmental policy process by using the complex method of risk management resulting in a less democratic decision-making process.

Role of Conflict in the Environmental Policy Process

Conflict can arise out of competition for the use of natural resources and the use of public lands by multiple partners (Vig & Kraft 2010). When individual incentives lead to environmental behaviors that have negative effects on society this leads to a resource dilemmas. Ignoring the effects personal behavior when using a public resource leads to the tragedy of the commons or the over-expenditure of natural resources by individuals in order to gain more for themselves exponentiation the effect by the number of users eventually the resource is exhausted, unless probably managed. Overgrazing on the open ranges of the Great Basin during the early

1900s was responsible for trampled soil and a lowered water table that lead to the great dust storms (EPA, 1994).

Conflict can also happen over legislation that is intended to protect rangelands from overgrazing, a practice that falls victim to the tragedy of the commons. The Bureau of Land Management (BLM) has the authority to manage, administer, and protect federal lands including regulating grazing under the Taylor Grazing Act of 1934 and the Federal Land Policy and Management Act of 1976 (ENS, 2007). On several occasions, the BLM has sued ranchers over grazing on public lands without a permit. Such was the case in Nevada where the U.S. Justice Department sued three ranchers, one posthumously, for multiple occurrences of deliberately grazing cattle on federally managed land (ENS, 2007).

When interested parties have differing opinions on the use of natural resources conflicts can arise such as over public lands that accommodate both wilderness and energy extraction (Vig & Kraft 2010). Public policies are concerned with how peoples' resource-use decisions influence the outcomes of resource dilemmas and thereby ought to be assessed in terms of both efficiency and equity.

Government agencies have an important role with respect to resource dilemmas making decisions that combine to explain the required, permitted and prohibited uses of natural resources (Vig & Kraft 2010).

Often the creation of hydroelectric dams cause the relocation of local residence creating conflict for the residents who find themselves forced off their lands. Such was the case in Sir Lanka where more than 111,400 families and about 700,000 people relocated in the months preceding April 1992 (Withanage, 1998). Additionally, the construction of the dam caused the destruction of forest increased soil erosion degradation of water quality downstream and river sedimentation. Moreover, the project failed at its promise of supplying sufficient water to the farmers (Withanage, 1998).

Collaborations can form out of conflicts such is the case in Washington where the State adopted a cooperative governance structure for Salmon recovery, which is a plan drafted by farmers, local governments, environmentalist, tribes and developers that will protect wetlands and flood plains, restore Salmon feeding grounds and retool hatcheries and dams in an attempt to restore Salmon Population (PCI, 2007).

Role of Sustainability in the Environmental Policy Process

Sustainable production and development are pressing issues in the debate over environmental policy. When individuals would rather choose a less optimal course of action one that typically satisfies some other highly-needs goal, a collective action dilemma occur (Vig and Kraft, 2010) instead of working collectively for a common goal that would benefit all the individuals. Business prefer to have no governmental requirements imposed on them to protect the environment rationalized by the businesses desire for maximized profits (Vig & Kraft, 2010). On the other hand, if laws set in place compelled the business to provide environmental protection, the businesses prefer approaches that allow for self-policing and regulatory flexibility (Vig & Kraft, 2010). As part of the collection action dilemma, the government chooses between a command-and control regulatory approach or a flexible and lenient approach, the latter of which could result in insufficient and slow change in industry toward sustainable production without strict regulation (Vig & Kraft, 2010).

To improve corporate environmental performance, government focus on approaches such as market incentives, self-reporting and environmental management system while limiting the command-and-control approach. The Environmental Protection Agency's Acid Rain Program is an approach using market-based incentives to control air pollution by allowing the trade of Sulfur Dioxide (SO_2). Under the program, a utility or industrial source must have enough allowances in their account that at a minimum must equal their annual sulfur dioxide emissions (EPA, 2009). According to the guidelines of the Acid Rain Program, allowances are commodities that may be bought, sold and traded by any private citizen, organization, corporation, municipality or interest groups (EPA, 2009).

Sustainability can be achieved by separating essential ideas in order to move toward sustainability. The first is to breakdown the notion that economic output comes from the increased use of energy and materials by, for instance, increasing tax on raw materials. This would encourage sustainable production innovations and altered consumer behavior (Vig and Kraft, 2010). Secondly, dissociating social well-being from gross domestic product (GDP) (essentially improving the quality of life faster than the rate at which wealth increases) through improved healthcare and/or education (Vig and Kraft, 2010). From a sustainability analysis it is incorrect to assume greater prosperity is the only way to improve well-being or to think an increase in the use of energy and raw materials is the exclusive path to increase prosperity.

The contrasting perspective, economism, recognizes prosperity as measured by GDP to be the overwhelming dominant goal of society and public policy (Vig and Kraft, 2010). According to Regional Environmental Center for Central and Eastern Europe, a city is sustainable when the current path of development does not take place at the expense of future generations and when equilibrium exists between different issues (REC, 2010).

The process of creating environmental policy receives influence from many parties all with an interest in the outcome of such policy decision. Market-based approaches, science, conflicts and sustainability all play a role in the process of creating environmental policy. All bring to the table positive attributes and as well as the negative; however, the voice of all are required to ensure the public's goods, their natural resources are protected and used in the most responsible manner.

References

Environmental News Service. (2007, August 30). U.S. sues property rights ranchers over grazing on federal lands. Retrieved from http://www.ens-newswire.com/ens/aug2007/2007-08-30-092.html

EurActiv. (2005). Risk-based policy-making. Retrieved from http://www.euractiv.com/en/science/risk-based-policy-making/article-133006

Environmental Protection Agency. (1994). Background for NEPA reviewers: Grazing on federal lands. Retrieved from http://www.epa.gov/compliance/resources/policies/nepa/grazing-federal-lands-pg.pdf

Environmental Protection Agency. (2009). Use of risk-based decision-making in UST corrective action programs OSWER Directive 9610.17 March 1, 1995. Retrieved from http://epa.gov/swerust1/directiv/od961017.htm#AttachmentA

Oregon Department of Transportation. (2010). House bill 2186 (HB 2186) metropolitan planning organization greenhouse gas emissions task force. Retrieved from http://www.oregon.gov/ODOT/TD/TP/HB2186.shtml

Pew Center on Global Climate Change. (2010). Workshop proceedings: Assessing the benefits of avoided climate change: Cost-benefit analysis and beyond. Retrieved from http://www.pewclimate.org/publications/report/workshop-proceedings-assessing-benefits-avoided-climate-change

Policy Consensus Initiative. (2007). Case study: Washington adopts collaborative governance structure for salmon recovery. Retrieved from http://www.policyconsensus.org/casestudies/docs/WA_sharedstrategy.pdf

Regional Environmental Center for Central and Eastern Europe. (2010). Main issues in a sustainable city. Retrieved from http://archive.rec.org/REC/Programs/Sustainablecities/Main.html

United States Environmental Protection Agency. (2009). Acid rain program SO_2 allowances fact sheet. Retrieved from. http://www.epa.gov/airmarkets/trading/factsheet.html

Withanage, H. (1998). Upper Kotmale hydropower project another disaster in dam history. *Centre for Environmental Justice.* Retrieved from http://www.ejustice.lk/CEJ%20web6/article%2023-%20Environment%20Conservation%20levy%20and%20climate%20adaptation.htm

The American Wilderness

The landscapes American call wildernesses are not pristine areas untouched by humankind as implied by the name. Rather, wilderness areas in American are creations of human influences from the Native Americans, the European settlers and from agricultural development. Human's reliance on nature to provide for their substance, generally is the reason for human alteration of the landscape.

Native Americans before the European settlers arrived relied on the land for substance. They hunted the wildlife and gather food instead of raising husbandry and developing farms. The Natives cleared the forest occasionally as a method of controlling the insect population and encouraging healthier growth of the vegetation. Although the influence the Natives had on the American landscape was minimal, it was an alteration; thereby invaliding the notion, the landscape represents rue definition of wilderness.

The European settlers when they arrived in American began clearing the forests for timber and for open land to farm. The settlers used the timber to build their homes and tools, heat their homes, and cook their food. As their demand for timber increased so did the clearing of the forests. The settlers utilized the cleared lands to garden. As the lands became deplete of their nutrients, the settlers began planting food on a new piece of land. On the former plot of land, faster growing trees and shrubs grew in place of the original forests. The alteration the European settlers caused to the American landscape was more dramatic than that caused by the Natives because their alteration involved replacing the native vegetation.

The most dramatic alteration of the American landscape occurred because of the growing population of the late 1800s/early 1900s. The demand for food for the growing urban populations created the drive to make new farms for agriculture. This resulted in the clearing of large portions of land in the Midwestern United States. The clearing of thousands of acres of forests to make farms caused not only the destruction of wildlife habitat it also caused erosion and flooding because of excess sentiment in the rivers. The necessity for the agricultural developments of the late 1800s/early 1900s is undeniable as is its impact on the American landscape.

Human alteration of the American landscape began before the arrival of the European settlers with the Natives who relied heavily on nature for substance. The European settlers continued the alteration, as they required more room for settlements. Moreover, the demand for agricultural resources by Americans growing population caused the most modification to the landscape. However, unrecognizable these occurrences of alteration are when observing the American wilderness they are undeniable. Therefore, to assume American wilderness areas are literally pristine is a mistake; rather they are definitely a creation of humanity.

Wilderness: The Effects of Human Expansion

The term "wilderness" may conjure images of forests, glades or marshes pristine in appearance, as if never touched by humanity. The modern person may use words like serene, beautiful and peaceful to describe wilderness. To the contrary, William Cronon (1996) explains in his article "The Trouble with Wilderness: Or, Getting Back to the Wrong Nature" that in the eighteenth century the word "wilderness" took its root from the Biblical meaning, referring to areas that were "savage," "barren" and a "waste" (18). The nineteenth century brought change to the idea of "wilderness" when it became a place associated with beauty and the divine, inspiring the idea of nature conservation; eventually leading to the designation of the first national park. By the beginning of the twentieth century, the idea of preservation over conservation became the leading theme of the protest lead by John Muir, founder of the Sierra Club over the Hetch Hetchy Dam who lost against the quest "reclaiming a wasteland…to put to human use" (Cronon 9) a decision reminiscent of the eighteenth century idea of wilderness. Secretary James R. Garfield in 1908 wrote in his explanation for approving the plans for the Hetch Hetchy Dam,

"Domestic use is the highest use to which water and available storage basins…can be put" (Nash, 1982). Charles Darwin wrote, "[who] would justify the extermination of the orangutan; so in literally thousands of cases would industrialist, politicians, and other public figures rationalize their efforts to turn the natural world to their own account" (Worster, 2006).

The modern American landscape underwent transformation because of human expansion and agriculture both of which required the extraction of natural resources that in the United States facilitated the creation of the Forest Service and the modern preservation movement. This land transformation came with consequences such as the displacement of native people, degradation of nature and conflict of land use. Cronon (1996) supposes many modern Americans view wilderness as places not yet touched by civilization stating, "Far from being the one place on earth that stands apart from humanity, it is quite profoundly a human creation – indeed, the creation of very particular human cultures at very particular moments in human history" (7). Proposing that instead of wilderness being the remaining remnant of an untouched land, it is the product of the civilization.

As early as 3,000 B.C., the European continental landscape experienced human expansion with the maximum expansion of the Alpines with Bronze Culture as recorded by Madison Grant in his article "The Passing of the Great Race" published in 1912 by the American Geographical Society. To begin with, Grant (1912) describes in anthropological terms how the Mediterranean race spread across Northern Africa and into Spain, Gaul and the British Isles and shows (with maps) how the Alpines spread out across central Europe westward and how the Nordic people spread across Northern Europe and the British Isles. Afterward there was "a great intrusion of Alpines [between the Mediterranean and the Nordic] throughout central Europe and thence expanding in every direction" (Grant, 1912 p.355). Subsequently, the Alpines, according to Grant (1912), taught the art of metallurgy to both the Mediterranean and the Nordic. Grant (1912) indicates the Nordic peoples gained such success with the science of metallurgy and such superior ability in the art of making and using weapons with their newfound abilities that they were able to turn against their conquerors and completely master them.

Grant (1912) continues with how the Nordics around the turn of the century tore apart the large and densely populated lands of the Alpines in Central Europe forcing them into dense populations scattered throughout the mountains and the infertile lands. The "submergence of the Alpines by the Nordics was so complete that their very existence was forgotten, until in our own day it was discovered that the central core of Europe was inhabited by a short, stocky, round-skull race, originally from Asia" (Grant, 1912, p. 357).

The displacement and eradication of the Alpines' from their native home lands by Nordic settlements parallels the actions taken in the United States against the Natives when the government "rounded up and moved [the Natives] onto reservations" (Cronon, 1996, p. 15). When creating wilderness areas such as Yellowstone, the United States Government needed to remove the Natives who had long used the area for substance fishing and hunting. There reason, to ensure the new tourist felt safe. "Preservation of the country's national parks and Indian removal proceeded in lock-step motion" (Steinberg, 2009, p.148). Ironically, Americans gained their symbolic wilderness, their given escape from their homes, work and life's troubles, from the forced eviction of the Indians. According to Cronon (1996) "The myth of the wilderness as 'virgin', uninhabited land had always been especially cruel when seen from the perspective of the Indians who had once called that land home," (p. 15) thus indicating how truly constructed the Nation's wildernesses are.

As late as 1100 A.D. the disappearance of the Mediterranean race, as Grant (1912) describes, in Central and Northern Europe is congruous with the Nordic and Slavic Alpines gaining supremacy in Spain and Northern Africa replacing the Alpine concentrations. The concentrations of the Alpines existed in two large centralized areas located in Eastern Europe, and with "the expansion of the…Nordics from Scandinavia and from the north of Germany is now at its maximum and they are everywhere pressing through the Empire of Rome and laying the foundation of the modern nations of Europe" (Grant, 1912, p. 358). The Nordic people secured for themselves the river valleys and fertile lands that were widely spread over large geographic areas where the "ruling, military aristocracy [were] more or less thinly scattered over a conquered peasantry of Mediterranean and Alpine blood" (Grant, 1912, p. 359) ultimately forcing the Alpines to living on infertile and mountainous regions.

Moreover, the behavior of American corporations during the seventeenth century parallels closely with what Grant describes happens between the Nordics and Alpines in Europe. Like the Nordics in Europe, American corporations secured for themselves the riverbanks and fertile lands of Cuba, Hawaii, the Philippines and Central America in a similar conquest for control over resources and the production of commodities (Tucker, 2007). "The world's greatest consumer economy [has] transformed not only American life and landscapes but distant places and societies as well" (Tucker, 2007, p. 217). Within the United States, exploitation of natural resources as commodities happened with little regard for the consequences and adverse effects on the environment. According to Nash (1982), "wherever, they [the pioneers] encountered wild county they viewed it through utilitarian spectacles: trees became lumber… and canyons the sites of hydroelectric dams" (p. 31). Economic, population, and urban growth were reasons for the utilization of natural resources and had a drastic effect on altering the landscape resulting in soil erosion, loss of animal habitat and species extinction.

Population growth, for example, spawns the demand for wood, which if unregulated can causes massive deforestation. In 1839, Americans consumed 1.6 billion board feet of lumber rising to 12.8 billion board feet by 1869 (Steinberg, 2009). Timber was not the only reason for clearing the forests; increasing urban populations left farmers looking for land to cultivate. According to Worster (2006) by 1700, over half a million acres of New England woodlands had been cleared for farming. Similarly, Cuba in a portion of its lands, by 1911, lost 15,000 acres of forest to agriculture rising to 43,000 acres (Tucker, 2007). Moreover, after the soil reached exhaustion farmers instead of amending the soil found it more convenient to have their servants clear more land (Steinberg, 2009). Such are examples of the historical events to prove Cronon's argument that the wilderness landscapes of today are a result of human influence over nature throughout history.

The consequences of deforestation, soil erosion and loss of animal habitats, were unknown at the time. An enduring effect of soil erosion began in 1780 when the port at Baltimore needed to be dredged on a regular basis to combat all the silt (Steinberg, 2009). Moreover, many animals, for instance the passenger pigeon, depended on the forest for their home; however, deforestation destroyed the passenger pigeon's habitat and nesting areas (Steinberg, 2009). Unfortunately, by the mid-nineteenth century, the loss of habitat in combination with over-hunting lead to the extinction of the passenger pigeon, an example of an extreme case Aldo Leopold termed "the impertinence of civilization" (Nash, 1982). Leopold once said, "[only] when the end of supply is in sight [do] we 'discover' that the thing is valuable (Nash, 1982).

Paul Sutter's (1998) essay "A blank spot on the map: Aldo Leopold, Wilderness and U. S. Service Recreational Policy" describes the circumstances surrounding how Leopold came to a decision in the early 1920s that the forest in the Western United States required a specific form of preservations. The newness of his ideas regarding wilderness and the qualities and reasons behind his resolutions combined with the "proliferation of the automobile, rampant road-building and a quantitative and qualitative boom in recreational demands made on public lands" (Sutter, 1998, p. 188) were the preoccupations of Leopold's wilderness advocacy. Leopold joined the Forest Service after graduating from Yale's Forestry School receiving his first post in 1909 with the Southwestern branch of the Forest Service at a time when the condition of the Forest Service system was failing to achieve Gifford Pinchot's dream of regulating private industry and by allowing its timber resources to deteriorate (Sutter, 1998). Nevertheless, monetary requirements and unpredictable changes in the market were reasons for the failure of the Forest Service to compete in the timber market combined with an adequate supply of timber form the private sector and diminishing market demand prevented the Forest Service from self-supporting itself (Sutter, 1998).

The Forest Service wilderness preservation plan restricted traditional local substance use of the lands such as grazing and timber extraction. The Forest Service and local residents clashed over, for instance the reduction of tax money received by the local governments because of the National Forest tax-exempt status or the new restrictions (Sutter, 1998). In the absence of serving the national timber market, foresters spent considerable time on local issues. As a solution to the growing concerns of the local citizens that the Forest Service share its revenue, access was granted to timber extraction for local consumption and a system of permits was set-up to accommodate other traditional substance uses; a system that would soon receive challenge from the introduction of a more perplexing national mission (Sutter, 1998). "The creation of the National Park Service in 1916 stands as an important watershed in the official recognition of scenic preservation as a legitimate form of land use" (Sutter, 1998, p. 192). By the mid-1910s, National Parks were scarce while National Forests were more extensive, close to population centers and offered hunting and fishing making them a more popular destination compared with the park systems that restricted visitor activity. With the emergence of car-camping came an increase in the reliance of public lands as a resting stop for travelers, and isolated place to hike or hunt all of which put more strain on the Forest Service (Sutter, 1998).

Leopold, according to Sutter (1998), was a "forester by training and a hunter-naturalist by avocation" (p. 204). Leopold was less interested in the in the scenic beauty of wilderness; he was a sportsman who appreciated the access modern roads provided to remote hunting locations while at the same time keenly aware of the dangers of overdevelopment. Leopold alleged development was "quickly delimiting the option for those who sought outing in areas free from such amenities [cabins, lodges, etc]" (Sutter, 1998, p. 204); an idea stemming from his entanglement with the forestry culture. Leopold defined wilderness as "a continuous stretch of country preserved in its original natural state, open to lawful hunting and fishing, big enough to absorb a two-week pack-trip and kept devoid of roads, artificial trails, cottages or other works of man" (Sutter, 1998, p. 207).

Leopold's new idea of protecting a frontier environment from the improvements of modern life such as improved roads and automobiles coincide with his vision of wilderness as a form of historical preservation. Leopold's vision reflects part of Cronon's reasoning for the modern transformation of the idea of wilderness as the sublime and the frontier. He suggests that Americans united the two ideas to remake wilderness in their own image, freighting it with moral values and cultural symbols that it carries to this day. One reason, the sublime steaming from the eighteenth century romantics who believed there were rare landscapes on Earth where one could glimpse the face of God; landscapes that would evoke emotions (Cronon, 1996). Another, the idea of frontier was a national myth that embodied the idea that wilderness as the place to return to a more primitive life allowing Americans a remedy for the troubles caused by the modern world. The latter of which was the force of Leopold's work at the Forest Service.

One of the qualities of wilderness that Leopold wanted to preserve was the limitation of road-developments and term permit developments in order to maintain at least one area remain free of roads leaving the reader to believe this was a major aspect of Leopold's desire to preserve wilderness (Sutter, 1998). Leopold wanted to preserve the public aspect of nature by restricting privatized forms of recreational development ensuring that large portions of the forest would remain open to all. Gifford Pinchot in 1910 wrote, the "natural resources of the Nation exist not for any small group, not for any individual, but for all the people" (Steinberg, 2009). Leopold was less concerned about the wildness of nature but rather on an individual travels through and lives within its confines. Leopold's intentions were to provide a form of recreation that would not consume its own resource base reasoning that wilderness was a "public recreational resource that needed no development" (Sutter, 1996, p. 213). Pinchot viewed conservation as "the development and use of the earth and all its resources for the enduring good of men" (Worster, 2006).

Sutter (1998) in his article describes Cronon as a critic of the modern idea of wilderness noting that Cronon views the purpose of the creation of the National Forest at the turn of the century as the preservation of "a pristine nature for American's leisure class" (p. 189). Leopold, Sutter (1998) indicates in the same article, shared Cronon's critique of the early twentieth century Americans' relationship with nature as dysfunctional because it was leisure based.

Cronon (1996) finds problem with the way Americans label wilderness agreeing that land should be set aside as wilderness. His problem is not with what Americans set aside but rather what they mean when they use the label wilderness and provided several paradoxes of what Americans label as wilderness (Cronon, 1996).

First the idea of biological diversity only originating in an untouched ecosystem as a possible solution to the protection of our environment while believing the only way to ensure the future of our ecosystem is through the most vigilant and self-conscious management (Cronon, 1996). Secondly, assuring our human intervention preservative of wilderness land does not cause, as it has in the past, the displacement of the native inhabitants thereby preventing paradoxical behavior.

Cronon (1996) explores solutions for how humans can hope to live naturally on the earth such as returning to a hunter-gather society and ensuring the balanced sustainable use of the land. Furthermore, he suggests persons should avoid separating the idea of wilderness from the trees planted in their yards by embracing the concept we make our home in nature. Cronon (1996) concludes that learning to honor the wild will cause a deeper reflection on how we use nature's resources believing strongly in achievement through "practicing remembrance and gratitude…[the] most basic of ways for us to recollect the nature, the culture, and the history that have come together to make the world as we know it" (p. 25).

Cronon's article "The Trouble with Wilderness: Or, Getting Back to the Wrong Nature", Grant's article "The Passing of the Great Race" and Sutter's article "A Blank Spot on the Map: Aldo Leopold, Wilderness and U.S. Forest Service Recreational Policy" describe the effects of human expansion, a theme present throughout each article. Grant wrote from the early 1900s offers insight into the expansion Europe experienced while Cronon and Sutter's essays written in the late 1900s offer insight into the preservation of wilderness in a large part due to human expansion. Additionally, disputes and conflict are an identifiable unifying theme between Cronon's, Grant's and Sutter's articles. Cronon in his article describes conflict with Natives over park construction; Grant describes clashes between the Nordic and Alpine races in Europe while Sutter's article describes conflict arising between the Forest Service and landowners. The outcomes of these disputes are ultimately the decisions that alter the landscape.

References

Cronon, W. (1996). "The Trouble with Wilderness: Or, Getting Back to the Wrong Nature." *Environmental History 1:1,* 7-28.

Grant, M. (1916). "The Passing of the Great Race". *Geographical Review 2:5,* 354-360.

Nash, R. F. (1982). *Wilderness and the Great American Mind.* New Have: Yale University Press.

Steinberg, T. *Down to Earth.* New York: Oxford University Press.

Tucker, R. P. (2007). Insatiable Appetite: The United States and the Ecological Degradation of the Tropical World. Lanham: Rowman & Littlefield Publishers.

Sutter, P. (1998). "A Blank Spot on the Map: Aldo Leopold, Wilderness and U.S. Forest Service Recreational Policy." *The Western Historical Quarterly.* Logan: Utah States University Press.

Worster, D. (2006). *Nature's Economy.* New York: Cambridge University Press.

Position Statement: Carbon Emissions

The United States should place limits on carbon dioxide (CO_2) emissions from the use of fossil fuels including a cap and trade system. Additionally, policymakers should pursue energy policies aimed at promoting and incentivizing the reduction of energy waste while simultaneously replacing subsidies to nonrenewable energy sources with subsidies to alternative energy source all the while considering the effects on the economy and the environment.

The cap and trade system sets a limit to the amount of allowable emissions resulting in reduced emissions at stationary sources. The system does not discourage economic activity, thus not hindering economic growth. Cap and trade system is working in the European Union, as has the Sulfur trade system in the United States. Furthermore, businesses have a financial incentive to reduce carbon emissions (Dr. Kaplan Yalcin, 2012).

Although not limits on emissions, incentives such as creating a voluntary carbon emissions reduction program through which participants receive tax credits, providing for natural sequestrate sites, promoting and subsidizing alternative energy sources such as solar, wind, hydroelectric and pebble-bed nuclear reactors are other methods through which the United States can achieve a reduction in carbon emissions.

Individuals and businesses could receive tax credits for reducing carbon emissions by decreasing energy waste and through restructuring projects. According to Daniel Schrag (2007), reducing energy consumption does not necessary mean consuming less, which might slow the economy instead he suggests "investing in low-energy adaptations such as efficient public transportation systems, or by adopting energy-efficient technologies in buildings, in automobiles, and throughout the economy (175)." Another technique to reducing carbon emissions is natural carbon sequestration. Government can encourage and reward, through tax incentives, agricultural and forestry practices improvements that provide for carbon sequestration (James Hansen et al., 2008). Additionally, the Intergovernmental Panel on Climate Change suggest policies provide financial incentives such as tax credits to agriculturalist for erosion control and protect soils. Moreover, the United States could eliminate subsidies to non-renewable energy provides and give subsidies to alternative energy providers. Wind, solar and hydroelectric energy sources are noncarbon emitting renewable energy sources that, if used, will reduce levels of carbon emissions.

Opponents to the suggestions above include Allison Macfarlane (2007) who states, "wind and solar energy…are intermittent and need to be supplemented by other secure forms of energy generation or by storage methods (168-169)." Moreover, Macfarlane (2007) contends that although hydroelectricity produces low cost energy it damages the environment through habitat destruction, community displacement, and interruption to the aquatic life cycle. Opposition to nuclear power refer to the problems of disposing high-level nuclear waste as well as the possibility of proliferation of nuclear weapons noting elimination of "carbon emissions from electricity generation by using nuclear power, for example, would require building two large nuclear plants each week for the next 100 years" (Macfarlane, 2007; Schrag, 2007).

In conclusion, the United States needs a healthy environment in which a strong economy can thrive. Currently, carbon emissions are threatening the health of the environment and it is my opinion government should step-up to combat said threat by implementing a cap and trade system.

References

EPA (2010) Inventory of U.S. Greenhouse Gas Emissions and Sinks: 1990-2008 Executive Summary, *United States Environmental Protection Agency* Figure ES-5 page 7 [online] http://www.epa.gov/climatechange/emissions/downloads10/US-GHG-Inventory-2010_ExecutiveSummary.pdf

Hansen, James, Makiko Sato, Pushker Kharecha, David Beerling, Robert Berner, Valerie Masson-Delmotte Mark Pagani, Maureen Raymo, Dana L. Royer and James C. Zachos (2008) Target Atmospheric CO2: Where Should Humanity Aim? *The Open Atmospheric Science Journal*, 2, 217-231 [online] http://my.oregonstate.edu/webapps/portal/frameset.jsp?tab_group=courses&url=%2Fwebapps%2Fblackboard%2Fcontent%2FcontentWrapper.jsp%3Fattachment%3Dtrue%26navItem%3Dcontent%26content_id%3D_2462765_1%26course_id%3D_234501_1%26displayName%3DHansen_James_350ppm.pdf%26href%3D%2F%2540%2540%2F3B4CE1773404 0F360E781B9279753535%2Fcourses%2F1%2FGEO_300_4

00_S2011%2Fcontent%2F_2462765_1%2FHansen_James_3 50ppm.pdf

Intergovernmental Panel on Climate Change. (2007). Climate Change 2007: Synthesis Report. *IPCC Fourth Assessment Report: Climate Change 2007* [online] http://www.ipcc.ch/publications_and_data/ar4/syr/en/spms4.html

Macfarlane, Allison M. (2008) Energy: The issue of the 21st Century. *Elements Vol 3,* pp 165-170 [online] http://my.oregonstate.edu/webapps/portal/frameset.jsp?tab_group=courses&url=%2Fwebapps%2Fblackboard%2Fcontent%2FcontentWrapper.jsp%3Fattachment%3Dtrue%26navItem%3Dcontent%26content_id%3D_2462765_1%26course_id%3D_234501_1%26displayName%3DMacfarlane_Energy-%2BIssue%2B21st%2BCentury.pdf%26href%3D%2F%2540%2540%2F3B4CE17734040F360E781B9279753535%2Fcourses%2F1%2FGEO_300_400_S2011%2Fcontent%2F_2462765_1%2FMacfarlane_Energy-%252520Issue%25252021st%252520Century.pdf

Schrag, Daniel P. (2007) Confronting the Climate-Energy Challenge. *Elements* Vol 3 pp 171-178) [online] http://my.oregonstate.edu/webapps/portal/frameset.jsp?tab_group=courses&url=%2Fwebapps%2Fblackboard%2Fcontent%2FcontentWrapper.jsp%3Fattachment%3Dtrue%26navItem%3Dcontent%26content_id%3D_2462765_1%26course_id%3D_234501_1%26displayName%3DSchrag_Confronting%2Bthe%2BClimate%2BEnergy%2BChallenge.pdf%26href%3D%2F%2540%2540%2F3B4CE17734040F360E781B9279753535%2Fcourses%2F1%2FGEO_300_400_S2011%2Fcontent%2F_2462765_1%2FSchrag_Confronting%252520the%252520Climate%252520Energy%252520Challenge.pdf

Yalcin, Dr. Kaplan. (2012) Lecture 15: Climate Change. *GEO 306 Minerals, water, energy, and the Environment*

Environmental Social Justice:
A Case Study of Northeast Portland, Oregon

Environmental social justice concerns arise out of situations of uneven distribution of environmental risks; such is the case in the northeast neighborhoods of Portland, Oregon where there is a large African American population (Dotterrer and Krishnan, 2011) as well as most of the Environmental Protection Agency's (EPA) hazard sites (EPA, n.d.). The hazards created by these sites affect, what Burger et al. (2001) describe as, communal property or resources, in this case air and water quality, used by the community and private property. Worldviews involved in this case are ecocentrism and contempocentrism held by neighborhood groups and industrialist, respectively. This paper examines the risks associated with environmental social justice in Portland, their uneven distribution, sources of the dissimilarities, reasons discrimination proof is difficult, and current actions attempting to fix the imbalance in environmental equality.

Uneven Distribution of Risks

Portland, Oregon is seventy-six percent White and six percent African American (Dotterrer and Krishnan, 2011) and the distribution of race is narrow with all of the highly concentrated African American communities situated within the Portland Neighborhood Associations' North Portland Neighborhood Services, Northeast Coalition of Neighborhoods, and the Central Northeast Neighborhoods districts (ONI, 2014). The narrow distribution of race leads to the uneven distribution of environmental hazard risks because most hazards sites are located within these neighbors.

Major Sources of Dissimilarity

Social, generational and procedural factors act as the major source of dissimilarity. In the neighborhoods of northeast Portland, social-economic factors of class and race contribute to environmental degradation. According to Dotterrer and Krishnan (2011), the 2005-2009 Median Household Income (MHI) for African American households was $26,988 to whereas the MHI for White households was $50,661 with thirty-two and a half percent of African Americans and roughly fourteen percent of Whites living below poverty illustrating a source of dissimilarity in the form of socio-economic imbalance.

Difficulties Proving Discrimination

Difficulty arises in trying to prove discrimination exists in the neighborhoods of northeast Portland because of vary perspective regarding the situation. For instance, it can be argued African American could choose to live in a different neighborhood, however, the cost of living may be higher in other neighborhoods, therefore not a viable chose. The low cost of rent may also be the reason industries choose to locate their business in these neighbors.

Action to Right the Wrong

The EPA (2014) is in action in northeast Portland righting the wrongs caused by industries, such as the clean-up occurring at the Portland Harbor Project Superfund site, which currently poses a threat to human exposure and the groundwater. Singer (2000) recommends the social relations model in which neighbors consider the consequences of their actions as they affect their neighbor when resolving conflicts amongst private property owners. Such is the actions of Oregon Humanities (2014), which hosts the Conservation Project, a public discussion program with forums, such as "White Out?: The Future of Racial Diversity in Oregon".

Conclusion

Environmental social justice depends on fair distribution of environmental risks, which often is not the case, such as the narrow distribution of race and hazardous sites in the neighborhoods of northeast Portland, Oregon. Everyone has a right to clean air and water regardless of socio-economic status. Resource managers, including urban designers, face a situation in which avoiding discrimination regarding distribution of environmental risk often can be seen as discrimination towards socio-economic groups, thus creating complicated urban planning decisions.

References

Burger, J., C. Field, R.B. Norgaard, E. Ostrom, and D. Policansky. (2001). "Common-pool resources and commons institutions: An overview of the applicability of the concept and approach to current environmental problems." In *Protecting the Commons: A Framework for Resource Management in the Americas*, edited by J. Burger, E. Ostrom, R.B. Norgaard, D. Policansky, and B.D. Goldstein, 1 – 15. Washington, DC: Island Press.

Environmental Protection Agency. (2014). Cleanup my community. Retrieved from http://ofmpub.epa.gov/apex/cimc/f?p=cimc:73::::71:P71_WELSEARCH:OR|State|OR|||true|true|true|true|true|true||-1|sites|N|basic

Environmental Protection Agency. (2014). Superfund site progress profile Portland Harbor. Retrieved from http://cumulis.epa.gov/supercpad/cursites/csitinfo.cfm?id=1002155

Office of Neighborhood Involvement. (2014). Portland Neighborhood Associations, District Coalitions & Offices with Boundaries Map. City of Portland. Retrieved from https://www.portlandoregon.gov/oni/article/283289

Oregon Humanities. (2014). Conversation Project. Retrieved from http://oregonhumanities.org/programs/conversation-project/

Oregon Humanities. (2014). White out? The future of racial diversity in Oregon. Retrieved from http://oregonhumanities.org/programs/2014-15-catalog/white-out/76/

Singer, J.W. (2000). *Property and Values: Alternatives to Public and Private Ownership*, edited by C. Geisler and G. Daneker, 3 – 19. Washington, DC: Island Press.

GLOSSARY

Cognitive approach: a way of dealing with something using mental action or process of acquiring knowledge and understanding through thought, experience, and the senses

Developing countries: a poor agricultural country that is seeking to become more advanced economically and socially

Dominant Resource Management Paradigm: current way of managing natural resources putting resource extraction above environmental protection

Ethnography: the scientific description of the customs of individual peoples and cultures

Environmental movement: a diverse scientific, social, and political movement for addressing environmental issues

Environmental interest groups: are generally public-interest groups, as their work benefits a wider community beyond their own active membership. They advocate for conservation and ecological issues.

Environmental policy: is the commitment of an organization or government to the laws, regulations, and other policy mechanisms concerning environmental issues

Environmental social justice: is the fair treatment and meaningful involvement of all people regardless of race, color, national origin, or income with respect to the development, implementation, and enforcement of environmental laws, regulations, and policies.

Globalization: the process by which businesses or other organizations develop international influence or start operating on an international scale.

Industrialization: the development of industries in a country or region on a wide scale.

Interest groups: a group of people that seeks to influence public policy on the basis of a particular common interest or concern

New Resource Management Paradigm: a new style of natural resource management that focuses on sustainability and preservation

Non-governmental organizations: groups independent of governments and international governmental organizations that are active in humanitarian, educational, health care, public policy, social, human rights, environmental, and other areas to affect changes according to their objectives

Public interest groups: groups pursuing goals the achievement of which ostensibly will provide benefits to the public at large, or at least to a broader population than the group's membership.

Population growth: the increase in the number of individuals in a population

Post-communist: the period of political and economic transformation or "transition" in former communist states located in parts of Europe and Asia in which new governments aimed to create free market-oriented capitalist economies.

Post-industrial: relating to an economy that no longer relies on heavy industry

Qualitative: relating to, measuring, or measured by the quality of something rather than its quantity

Quantitative: relating to, measuring, or measured by the quantity of something rather than its quality

Sustainability: avoidance of the depletion of natural resources in order to maintain an ecological balance

Wilderness: an uncultivated, uninhabited, and inhospitable region

INDEX

Australia 23, 30, 31, 32, 35, 36, 37, 40, 41, 42

Buddhist 25, 32, 33

Carbon 17, 63, 79, 111, 112, 113

Christianity 33, 34

Conflict 15, 39, 56, 73, 74, 79, 80, 82, 83, 84, 87, 96, 109, 120

Congress 40, 35, 69, 70, 73, 74

Cognitive approaches 8, 9

Developing countries 38, 43

Dominant Resource Management Paradigm 15, 18

Economic 37, 38, 45, 47, 48, 49, 50, 51, 53, 73, 75, 79, 86, 111, 118, 119, 120

Ethnography 9

Environmental Hypocrites 19

Environmental movement 13

Environmental interest groups 63, 67

Environmental policy 5, 15, 16, 18, 22, 23, 25, 27, 29, 30, 31, 32, 33, 34, 35, 37, 39, 42, 44, 45, 57, 58, 59, 63, 72, 73, 74, 79, 80, 82, 85, 87, 88

Environmental social justice 117, 120

Globalization 49, 53, 54, 61, 62

Hindu 25, 32, 33

Industrial 24, 26, 29, 38, 39, 44, 50, 53, 63, 68, 69, 71, 86

Industrialization 26, 45, 51

Industry 56, 65, 68, 69, 70, 73, 75, 77, 81, 85, 103

Interest groups 65, 66, 67, 70, 71, 75, 77, 82, 86

Institutional 37, 44, 56, 57, 60

Islam 33, 34

Judaism 33

Korea, 35, 36

Markets 54, 79

Nepal 30, 35, 36

New Resource Management Paradigm 13, 15, 18

Non-governmental organizations 48, 57, 63, 64, 65, 75, 82

President 17, 27, 40, 42, 72, 73, 77

Public interest groups 65, 66, 67, 75

Population Growth 49, 50, 61, 62, 101

Post-communist 43, 45, 46

Post-industrial 13, 15, 37, 40, 44, 45, 46, 47, 49, 50, 51, 53, 58, 64

Qualitative 8, 9, 10, 102

Quantitative 8, 9, 102

Quota sampling 10

Rural 13, 14, 15, 17, 18, 21, 65

Science 44, 45, 48, 55, 56, 58, 59, 62, 77, 79, 80, 87, 89, 97, 114

Sustainability 15, 39, 56, 73, 74, 79, 80, 82, 83, 84, 87, 96, 109, 120

United States 35, 37, 40, 41, 44, 46, 63, 64, 70, 72, 77, 91, 93, 96, 98, 100, 102, 110, 111, 112, 113

Urban 13, 14, 15, 17, 18, 21, 24, 26, 30, 49, 50, 51, 53, 61, 62, 64, 93, 100, 101, 120

Value Change 13, 49, 53

Wilderness 64, 65, 83, 92, 94, 95, 96, 98, 99, 101, 102, 103, 105, 106, 107, 108, 109, 110

www.ingramcontent.com/pod-product-compliance
Lightning Source LLC
Chambersburg PA
CBHW030659220526
45463CB00005B/1847